YOUR KNOWLEDGE HAS VALUE

- We will publish your bachelor's and
  master's thesis, essays and papers

- Your own eBook and book -
  sold worldwide in all relevant shops

- Earn money with each sale

Upload your text at www.GRIN.com
and publish for free

Bibliographic information published by the German National Library:

The German National Library lists this publication in the National Bibliography; detailed bibliographic data are available on the Internet at http://dnb.dnb.de .

Imprint:

Copyright © 2016 GRIN Verlag
Print and binding: Books on Demand GmbH, Norderstedt Germany
ISBN: 9783668682672

This book at GRIN:

https://www.grin.com/document/387601

Vidhya Ganesh Rangarajan

Effectiveness of Contrast Limited Adaptive Histogram Equalization on Multispectral Satellite Imagery

GRIN Verlag

EFFECTIVENESS OF CONTRAST LIMITED ADAPTIVE HISTOGRAM EQUALIZATION TECHNIQUE ON MULTISPECTRAL SATELLITE IMAGERY

A Report

Submitted in fulfilment of the requirements of

NITK SUMMER INTERNSHIP PROGRAMME – 2016

by

VIDHYA GANESH R

(MT/RS/10001/2015)

Department of Remote Sensing

Birla Institute of Technology, Mesra, Ranchi – 835215

DEPARTMENT OF APPLIED MECHANICS AND HYDRAULICS

NATIONAL INSTITUTE OF TECHNOLOGY KARNATAKA

SURATHKAL, MANGALORE – 575025

JULY 2016

ACKNOWLEDGEMENT

I express my deep gratitude to **Dr. H. Ramesh**, Assistant Professor, Department of Applied Mechanics and Hydraulics, National Institute of Technology Karnataka, Surathkal for his valuable guidance, sportive encouragement, and valuable suggestions throughout my internship. I wish to put on record my deep reverence and esteemed thanks to him.

I express my sincere thanks to **Dr. G. S. Dwarakish**, Professor and Head, Department of Applied Mechanics and Hydraulics, National Institute of Technology Karnataka, Surathkal for extending the facilities in the department throughout my dissertation work. I am also thankful to **Dr. A. P. Krishna**, Professor and Head, Department of Remote Sensing, Birla Institute of Technology, Mesra for giving me the permission to pursue my summer internship from National Institute of Technology Karnataka, Surathkal.

My special thanks to all my **friends** in NITK for their everlasting support and encouragement throughout my internship. I express earnest thanks to all the **teaching** and **non-teaching staff**, Department of Applied Mechanics and Hydraulics, National Institute of Technology Karnataka for their direct or indirect help during the course of my internship.

I also express my deep gratitude to my beloved **parents** and **sister** for their moral support and constant encouragement throughout my internship.

VIDHYA GANESH R

ABSTRACT

Contrast Limited Adaptive Histogram Equalization technique (CLAHE) is a widely used form of contrast enhancement, used predominantly in enhancing medical imagery like X-rays and to enhance features in ordinary photographs.

This work is aimed to understand the effectiveness of using this technique in multispectral satellite imagery and to study its effectiveness in different regions of the electromagnetic spectrum. This work also aimed in analysing variations of spatial and spectral resolutions of a sensor affect the performance of the CLAHE technique by means of comparing quantitative parameters of the enhanced images between the sensors.

A new performance parameter called Degree of Contrast Enhancement (DCE) has also been formulated so as to quantify the amount of increase/decrease in contrast between the enhanced and original images on application of the CLAHE algorithm on it.

A general idea of the feature that can be enhanced in each spectral region was also studied. The results showed that the technique was most effective for shorter wavelengths when compared to longer wavelength regions.

A comparative study between the CLAHE technique and the conventional global histogram equalization technique resulted in the former technique emerging superior of the two and thereby reconstructed images of better quality.

CONTENTS

LIST OF TABLES

CHAPTER 1 - INTRODUCTION

Image enhancement is among the simplest and most appealing areas of digital image processing. Basically, the idea behind enhancement techniques is to bring out detail that is obscured, or simply to highlight certain features of interest in an image. It is important to keep in mind that enhancement is a very subjective area of image processing. Improvement in quality of degraded images can be achieved by using application of certain enhancement techniques [7].

Image enhancement for satellite imagery can be classified into three types [10] [11]

a) Spatial Enhancement (Filtering),

b) Spectral Enhancement (Band ratio-ing) and

c) Radiometric Enhancement (Contrast Enhancement)

The present study is dealing with the radiometric enhancement of the satellite imagery by applying Contrast Limited Adaptive Histogram Equalization technique. This technique is extensively used in enhancement of medical images like X-rays of various body organs and is found to give satisfactory results in that domain [6].

Adaptive histogram equalization (AHE) is a computer image processing technique used to improve contrast in images. It differs from ordinary histogram equalization in the respect that the adaptive method computes several histograms, each corresponding to a distinct section of the image, and uses them to redistribute the lightness values of the image. It is therefore suitable for improving the local contrast and enhancing the definitions of edges in each region of an image.

However, AHE has a tendency to over amplify noise in relatively homogeneous regions of an image. A variant of adaptive histogram equalization called contrast limited adaptive histogram equalization (CLAHE) prevents this by limiting the amplification [7]. Ordinary histogram equalization uses the same transformation derived from the image histogram to transform all pixels. This works well when the distribution of pixel values is similar throughout the image. However, when the image contains regions that are significantly lighter or darker than most of the image, the contrast in those regions will not be sufficiently enhanced.

1.1 Global Histogram Equalization

An alternative technique for contrast enhancement which has been widely used is global histogram equalization [13]. In this method, the intensity values in the image are altered such that the resulting image has a constant intensity histogram. This transformation may be accomplished by the use of the cumulative distribution function of the pixel intensities as the intensity remapping function. Such images utilize the available display levels well, but because the contrast enhancement is based on the statistics of the entire image, some levels will be used for the depiction of parts of the image which are diagnostically unimportant, such as the background of the image.

1.2 Adaptive Histogram Equalization

Adaptive histogram equalization (AHE) developed by Ketcham et al, Hummel, Pizer et al [14] [15] [16] improves on this by transforming each pixel with a transformation function derived from a neighbourhood region. It was first developed for use in aircraft cockpit displays [15] [17]. In its simplest form, each pixel is transformed based on the histogram of a square surrounding the pixel, as in shown Fig 1. The derivation of the transformation functions from the histograms is exactly the same as for ordinary histogram equalization. The transformation function is proportional to the cumulative distribution function (CDF) of pixel values in the neighbourhood [18].

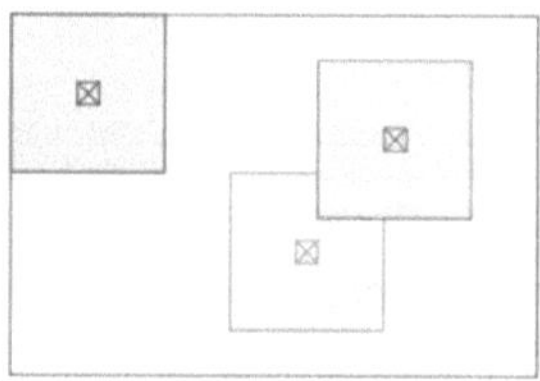

Fig 1.1 Concept of Adaptive Histogram Equalization

Pixels near the image boundary have to be treated specially, because their neighbourhood would not lie completely within the image. This applies for example to the pixels to the left or above the blue pixel in Fig 1.1. This can be solved by extending the image by mirroring pixel lines and columns with respect to the image boundary. Simply copying the pixel lines on the border is not appropriate, as it would lead to a highly peaked neighbourhood histogram.

1.3 Contrast Limited Adaptive Histogram Equalization

Contrast Limited AHE (CLAHE) developed by Pizer et al [19] [20], differs from ordinary adaptive histogram equalization in its contrast limiting [6]. This feature can also be applied to global histogram equalization, giving rise to contrast limited histogram equalization (CLHE), which is rarely used in practice.

In the case of CLAHE, the contrast limiting procedure has to be applied for each neighbourhood from which a transformation function is derived. CLAHE was developed to prevent the over-amplification of noise that adaptive histogram equalization can give rise to. This is achieved by limiting the contrast enhancement of AHE. The contrast amplification in the vicinity of a given pixel value is given by the slope of the transformation function. This is proportional to the slope of the neighbourhood cumulative distribution function (CDF) and therefore to the value of the histogram at that pixel value. CLAHE limits the amplification by clipping the histogram at a predefined value before computing the CDF. This limits the slope of the CDF and therefore of the transformation function. The value at which the histogram is clipped, the so-called clip limit, depends on the normalization of the histogram and thereby on the size of the neighbourhood region. Common values limit the resulting amplification is between 3 and 4. It is advantageous not to discard the part of the histogram that exceeds the clip limit but to redistribute it equally among all histogram bins (Fig 1.2).

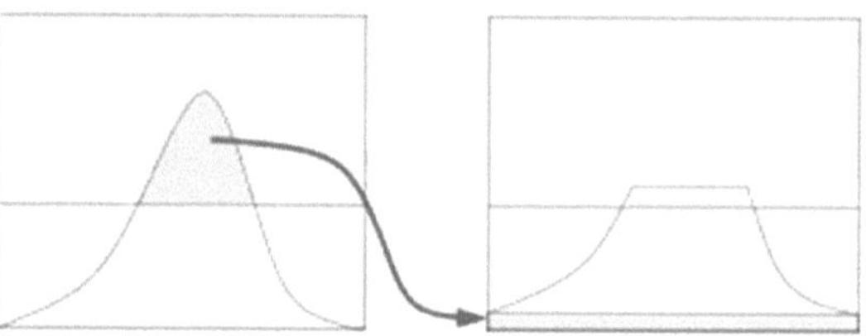

Fig 1.2 Redistribution of pixels above clip limit in Contrast Limited Adaptive Histogram Equalization

The redistribution will push some bins over the clip limit again (region shaded green in the Fig 1.2), resulting in an effective clip limit that is larger than the prescribed limit and the exact value of which depends on the image. If this is undesirable, the redistribution procedure can be repeated recursively until the excess is negligible. This technique has not yet been applied to satellite imagery which are large in size and contain various features that need to be distinguished from each other accurately. Hence the use of this technique in the geospatial domain is studied.

Algorithm Steps [7]:

1. Obtain all the inputs: Image, Number of regions in row and column directions, Number of bins for the histograms used in building image transform function (dynamic range), Clip limit for contrast limiting (normalized from 0 to 1)
2. Pre-process the inputs: Determine real clip limit from the normalized value if necessary, pad the image before splitting it into regions
3. Process each contextual region (tile) thus producing gray level mappings: Extract a single image region, make a histogram for this region using the specified number of bins, clip the histogram using clip limit, create a mapping (transformation function) for this region
4. Interpolate gray level mappings in order to assemble final CLAHE image: Extract cluster of four neighbouring mapping functions, process image region partly overlapping each of the mapping tiles, extract a single pixel, apply four mappings

to that pixel, and interpolate between the results to obtain the output pixel; repeat over the entire image

```
Syntax: J = adapthisteq(I,param1,val1,param2,val2...)
```

`adapthisteq(I)` enhances the contrast of the grayscale image I by transforming the values using contrast-limited adaptive histogram equalization (CLAHE) in MATLAB R2013a® [21] [22]. CLAHE operates on small regions in the image, called tiles, rather than the entire image. Each tile's contrast is enhanced, so that the histogram of the output region approximately matches the histogram specified by the `'Distribution'` parameter. The neighbouring tiles are then combined using bilinear interpolation to eliminate artificially induced boundaries. The contrast, especially in homogeneous areas, can be limited to avoid amplifying any noise that might be present in the image.

Parameter names can be abbreviated, and case does not matter. The additional parameters include 'NumTiles' for specifying the number of tiles by row and column with the default set at [8 8], 'ClipLimit' to define the limit of clipping the histogram, 'NBins' to define the number of bins in the histogram, 'Distribution' to define the distribution of the histogram – either 'uniform' for a flat histogram, 'rayleigh' for a bell shaped histogram or 'exponential' [22].

CHAPTER 2 - DATA ACQUISITION

The area under study covers the towns of Hejamadi, Karnad and Nadsal, of Mangalore district which lie along the Malabar coast, enroute Mangalore to Udupi. This area contains a variety of land cover and land use features. The Arabian Sea lies towards the west of the mainland and the Shambhavi river flows through the town Hejamadi.

Datasets from four sensors of different make and spatial resolutions have been obtained, namely the Advanced Wide-Field Sensor (AWiFS) (56m) and Linear Imaging Self-Scanning Camera (LISS III) (23.5m) from the IRS Resourcesat-I mission, Operational Land Imager (OLI) (30m) from the Landsat 8 mission and Multi-Spectral Instrument (MSI) (10m in visible and 20m in SWIR region) from the Sentinel 2A mission.

The details of the datasets collected and used for the present study are given in Table 2.1

Table 2.1 Details of the datasets used for the study

Sensors	Mission	Date of Acquisition	Spatial Resolution (m)	Spectral Resolution (Central Band Wavelength) (μm)			
				Green	Red	NIR	SWIR
LISS III	Resourcesat-I	23-01-2013	23.5	0.555	0.65	0.815	1.625
AWiFS	Resourcesat-I	09-05-2013	56	0.555	0.65	0.815	1.625
OLI	Landsat 8	02-09-2013	30	0.56	0.655	0.865	1.610
MSI	Sentinel 2A	22-05-2016	10m (Visible) 20m (SWIR)	0.56	0.665	0.842	1.610

The visible and IR bands of the satellite imagery were used for the study and methodology shown in Fig.3.1 is adopted. The images were enhanced using the CLAHE technique in MATLAB R2013a®. The technique was applied to enhance the green, red, NIR and SWIR bands of all the four datasets separately. The single band images were taken as the input images in MATLAB and the adaptive histogram equalization function was used over the image. The distribution of the neighbourhood was assumed to take the form of the Rayleigh distribution and a clip limit of 0.01 was used for all the images. The effectiveness of the enhancement technique was evaluated using two performance parameters such as Mean Square Error (MSE), Peak Signal-to-Noise Ratio (PSNR) and Degree of Contrast Enhancement (DCE).

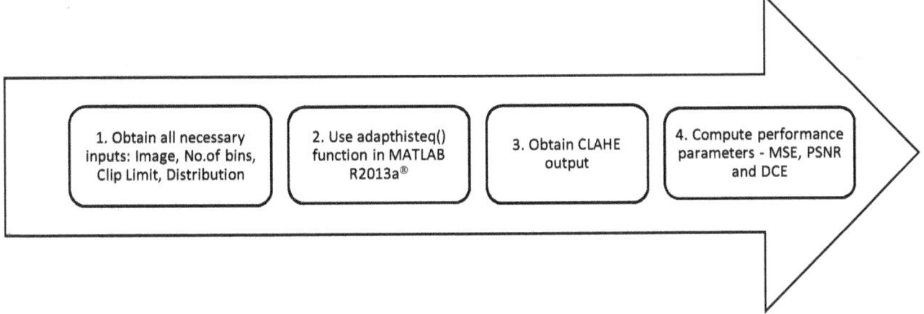

Fig 3.1 Flowchart of the methodology adopted

3.1 Mean Square Error (MSE)

MSE is the simplest and most widely used quality metric and is computed by averaging the squared intensity differences of original image and enhanced image [3]. The mean squared error (MSE) for all practical purposes allows us to compare the "true" pixel values of original image to degraded image. The MSE represents the average of the squares of the "errors" between actual image and noisy image. The error is the amount by which the values of the original image differ from the degraded image. Mean square error (MSE) is given by Eqn. (1).

$$MSE = \frac{\Sigma((f(i,j) - F(i,j))^2}{MN} \qquad (1)$$

where,

f (i, j) is the original image, F (i, j) is the enhanced image and M *N is the size of image with respect to rows and columns.

3.2 Peak Signal to Noise Ratio (PSNR)

PSNR is used to calculate the ratio between the maximum possible power of a signal and the power of corrupting noise that affects the fidelity of its representation [3][1][4]. The term peak signal-to-noise ratio (PSNR) is an expression for the ratio between the maximum possible value (power) of a signal and the power of distorting noise that affects the quality of its representation. Because many signals have very wide dynamic range, (ratio between the largest and smallest possible values of a changeable quantity) the PSNR is usually expressed in terms of the logarithmic decibel scale.

Image enhancement or improving the visual quality of a digital image can be subjective. Saying that one method provides a better quality image could vary from person to person. For this reason, it is necessary to establish quantitative/empirical measures to compare the effects of image enhancement algorithms on image quality. Using the same set of tests images, different image enhancement algorithms can be compared systematically to identify whether a particular algorithm produces better results or not. The metric under investigation is the peak-signal-to-noise ratio. If we can show that an algorithm or set of algorithms can enhance a degraded known image to more closely resemble the original, then we can more accurately conclude that it is a better algorithm.

The Peak Signal-to-Noise Ratio (PSNR) is given by Eqn. (2).

$$PSNR = 10log_{10}(\frac{R^2}{MSE}) \qquad\qquad (2)$$

where,

R is the maximum signal value that can exist in our original image (i.e., 65535 owing to consideration of satellite images having radiometric resolution 16-bit), MSE is the mean square error as computed previously.

3.3 Degree of Contrast Enhancement (DCE)

Mean Square Error and Peak Signal-to-Noise Ratio are effective measures used to evaluate how well the image is reconstructed after applying a particular algorithm or contrast enhancement technique. They measure how well the enhanced image matches the original image.

However, they cannot judge how much of contrast has been enhanced due to the application of the contrast enhancement algorithm. Hence, a new performance parameter called Degree of Contrast Enhancement (DCE) has been formulated. This involves determination of the magnitude of increase/decrease in contrast ratio between the enhanced and the original images and representing the same as a percentage of the contrast of the input image. DCE is given by Eqn. (3)

$$DCE\ (\%) = \frac{|Contrast\ Ratio\ of\ Enhanced\ Image - Contrast\ Ratio\ of\ Input\ Image|}{Contrast\ Ratio\ of\ Input\ Image} x100 \qquad (3)$$

where,

$$Contrast\ Ratio = \frac{Maximum\ Brightness\ Value\ in\ the\ Image}{Minimum\ Brightness\ Value\ in\ the\ Image} \qquad (4)$$

On applying CLAHE for different bands in the LISS III image, the following results were obtained (Fig 4.1).

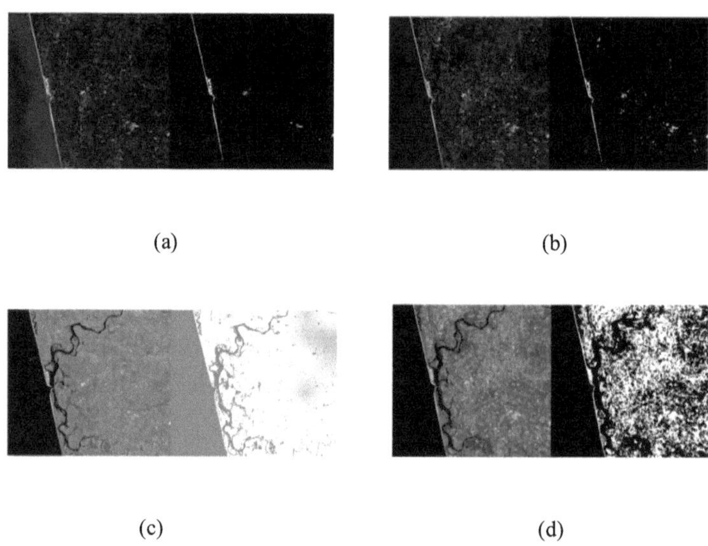

(a)   (b)

(c)   (d)

Fig 4.1 Original and enhanced images of the study area in LISS III in (a) Green (b)Red (c) NIR and (d) SWIR bands

The shoreline was significantly enhanced in the lower wavelengths (green and red), blackening out all other features. In the NIR band, the major water body (river) was enhanced prominently in a darker tone, owing to low spectral reflectance of water in the IR region, whitening out all other features. In the SWIR band, major water bodies and wetlands were enhanced in a darker tone, though the enhancement was ambiguous and not distinct.

The results of the technique on the AWiFS image were as follows (Fig 4.2).

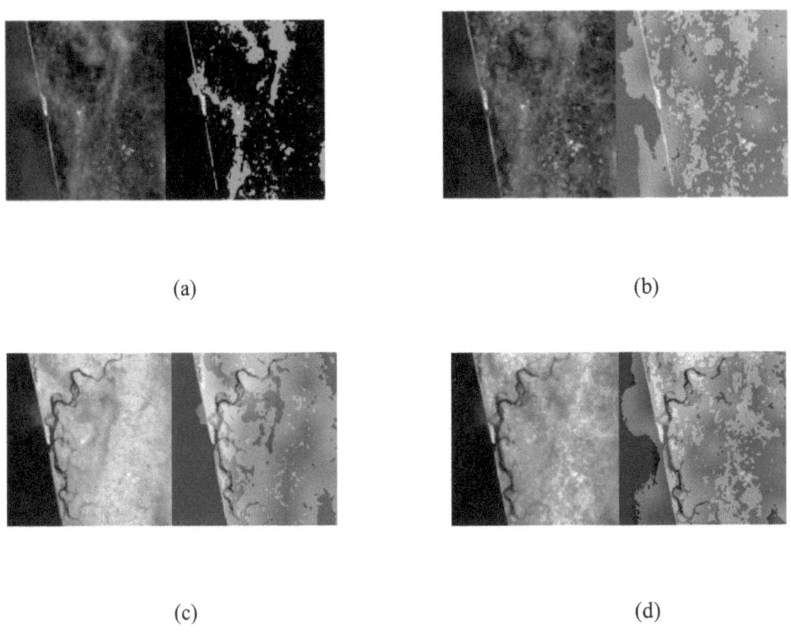

(a)   (b)

(c)   (d)

Fig 4.2 Original and enhanced images of the study area in AWiFS in (a) Green (b)Red
(c) NIR and (d) SWIR bands

In the AWiFS images, the application of CLAHE on the green and red bands show no significant feature enhancement. However, the wetlands and water bodies are enhanced in a darker tone in the NIR band. In the SWIR band, the enhancement is ambiguous and not distinct for a particular feature. Overall, application of the technique in the AWiFS image was not found to be that productive in feature enhancement or extraction.

On applying CLAHE for different bands in the Landsat 8 OLI image, the following results were obtained (Fig. 4.3).

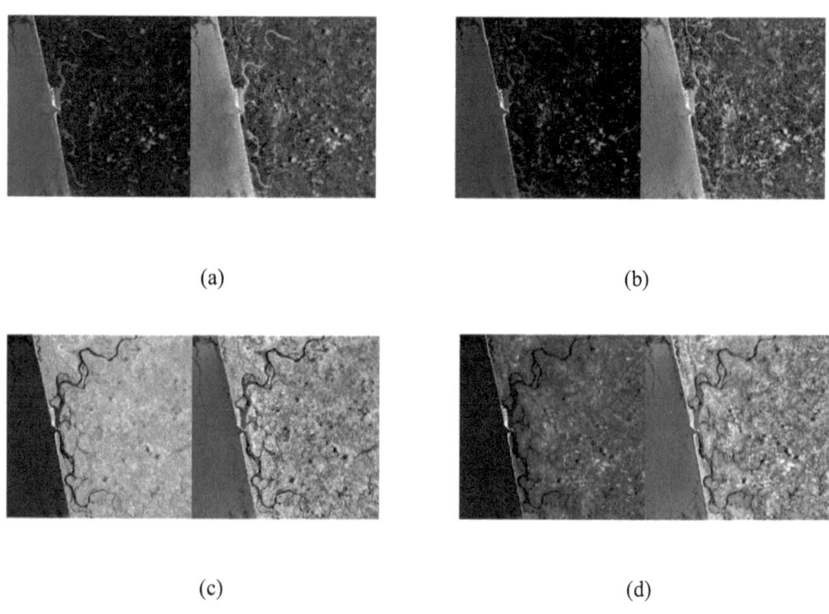

(a)

(b)

(c)

(d)

Fig 4.3 Original and enhanced images of the study area in Landsat 8 OLI in (a) Green (b)Red (c) NIR and (d) SWIR bands

Wave ripples are greatly enhanced in the Landsat images, especially in the lower wavelengths, owing to its greater spatial resolution and narrower bandwidth. Land features are better visible in the lower wavelength regions, although no particular feature is enhanced separately. River, inland water bodies and wetlands were enhanced significantly in a darker tone in the NIR band. River portions having greater sediments may be enhanced, leading to multiple tones in the river. However, interpretation of sediment waters is ambiguous and needs to be accompanied by associated ground truthing results. In the SWIR band, only clouds and their shadows are enhanced to an extent.

Sentinel 2A is by far, the highest resolution freely available multispectral dataset available to end-users. On applying CLAHE for different bands in the Sentinel 2A MSI image, the following results were obtained (Fig 4.4).

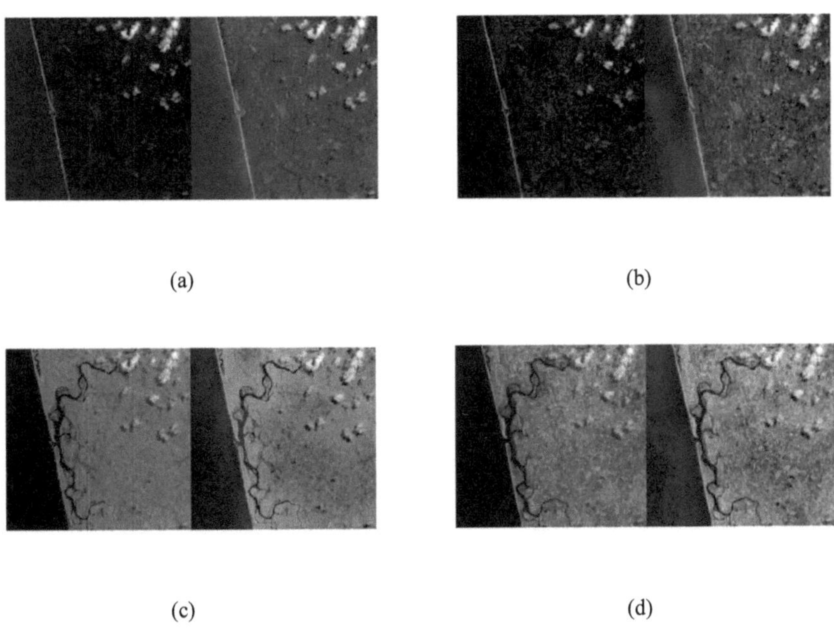

(a)   (b)

(c)   (d)

Fig 4.4 Original and enhanced images of the study area in Sentinel 2A MSI in (a) Green (b)Red (c) NIR and (d) SWIR bands

No significant enhancement of a particular feature was found in the green and red bands. The river was enhanced significantly in the NIR band in a darker tone. The technique however provided great results in the SWIR band. There were prominent enhancement of clouds and their shadows. Wetlands were also prominently enhanced. River portions having greater sediments may be enhanced, leading to multiple tones in the river. However, interpretation of sediment waters is ambiguous and needs to be accompanied by associated ground truthing results.

So as to perform a comparative study as to which among the CLAHE and global histogram equalization techniques is better, the AWiFS dataset was taken and global histogram equalization was performed on its four bands and the quality parameters were computed.

CHAPTER 5 - RESULTS AND INFERENCES

The judgement parameters, namely the Mean Square Error (MSE), Peak Signal-to-Noise Ratio (PSNR) and Degree of Contrast Enhancement (DCE) were computed for the resultant enhanced image of each band of each dataset and were compared. The resultant values for the **Green Band** in the datasets acquired were shown in Table 5.1.

Table 5.1 Observations for the Green Band in the four datasets

	MSI	LISS III	OLI	AWiFS
Wavelength(μm)	0.560	0.555	0.560	0.555
Resolution(m)	10	23.5	30	56
No.of pixels	12,59,356	2,06,769	1,40,178	52,140
MSE	6.325×10^7	3.1379×10^7	5.8417×10^7	4.2124×10^7
RMSE	7.953×10^3	5.6017×10^3	7.6431×10^3	6.4903×10^3
PSNR(dB)	18.3188	21.3631	18.6641	20.0842
DCE (%)	6.1	59.9	11.5	41.5

Similarly, the resultant values for the **Red Band** in the datasets acquired were presented in Table 5.2.

Table 5.2 Observations for the Red Band in the four datasets

	MSI	LISS III	OLI	AWiFS
Wavelength(μm)	0.665	0.65	0.655	0.65
Resolution(m)	10	23.5	30	56
No.of pixels	12,59,356	2,07,680	1,40,178	52,140
MSE	7.2895×10^7	3.1684×10^7	6.8536×10^7	4.964×10^7
RMSE	8.5378×10^3	5.6289×10^3	8.2786×10^3	7.0455×10^3
PSNR(dB)	17.7025	21.3210	17.9703	19.3712
DCE (%)	32.5	82.5	2.8	28.8

The resultant values for the **NIR Band** in the datasets acquired were presented in Table 5.3.

Table 5.3 Observations for the NIR Band in the four datasets

	MSI	LISS III	OLI	AWiFS
Wavelength(μm)	0.842	0.815	0.865	0.815
Resolution(m)	10	23.5	30	56
No.of pixels	12,59,356	2,07,680	1,40,178	52,140
MSE	11.060x10⁷	5.0819x10⁷	9.2603x10⁷	5.8267x10⁷
RMSE	10.517x10³	7.1287x10³	9.6230x10³	7.6333x10³
PSNR(dB)	15.8918	19.2692	16.6632	18.6753
DCE (%)	56.4	83.3	113.2	63.8

The resultant values for the **SWIR Band** in the datasets acquired were shown in Table 5.4.

Table 5.4 Observations for the SWIR Band in the four datasets

	MSI	LISS III	OLI	AWiFS
Wavelength(μm)	1.610	1.625	1.610	1.625
Resolution(m)	20	23.5	30	56
No.of pixels	3,15,675	2,07,680	1,40,178	52,140
MSE	12.257x10⁷	4.1813x10⁷	10.440x10⁷	5.6951x10⁷
RMSE	11.071x10³	6.4663x10³	10.218x10³	7.5466x10³
PSNR(dB)	15.4455	20.1164	16.1423	18.7744
DCE (%)	67.2	92.4	18.1	77.0

By comparing the MSE and PSNR of the four datasets in each band, it can thus be inferred that no distinct relationship can be generated between the efficiency of the CLAHE technique and the spatial resolution of the datasets alone. However, that is not the case with respect to the spectral resolution of the datasets. The variation of MSE (Fig 5.1), PSNR (Fig 5.2) and DCE (Fig 5.3) are plotted with respect to wavelength.

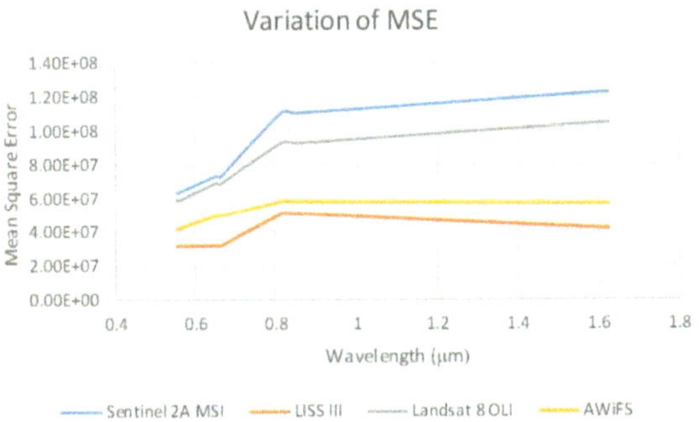

Fig 5.1 Variation of Mean Square Error (MSE) with respect to wavelength

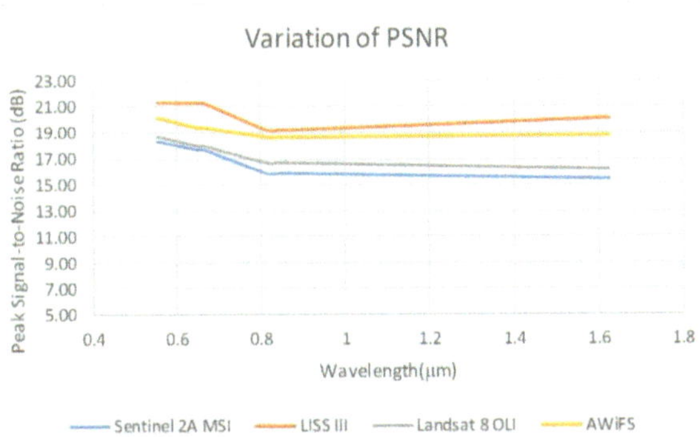

Fig 5.2 Variation of Peak Signal-to-Noise Ratio (PSNR) with respect to wavelength

The higher the PSNR, the better the degraded image has been reconstructed to match the original image and the better the reconstructive algorithm [7]. This would occur because we wish to minimize the MSE between images with respect the maximum signal value of the image.

We see that the Mean Square Error (MSE) is lower for shorter wavelengths of the electromagnetic spectrum and keeps increasing for greater wavelengths (Fig 5.1) thereby decreasing the Peak Signal-to-Noise Ratio (PSNR) (Fig 5.2) and the suitability of using this method on a whole on the image.

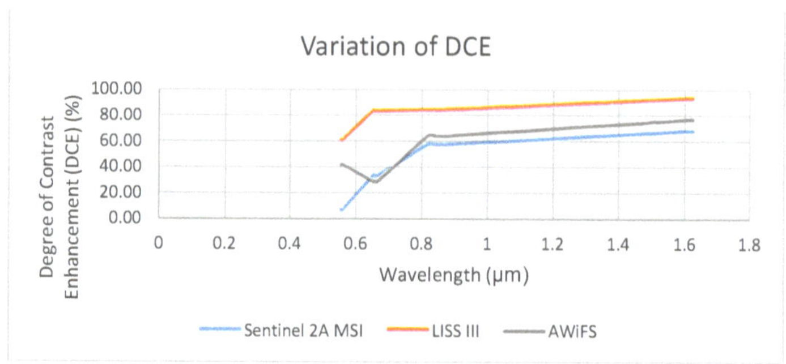

Fig 5.3 Variation of Degree of Contrast Enhancement (DCE) with respect to wavelength

Although PSNR measures how effective the image reconstruction process is after the algorithm has been applied, the degree of contrast enhancement (DCE) defines the amount by which contrast has been enhanced in each band.

It is seen that the DCE increases with respect to wavelength thereby providing a view that **images are better reconstructed but are of poorer contrast at shorter wavelengths** and vice versa at longer wavelengths of the spectrum.

The comparative values of MSE and PSNR recorded when Global Histogram Equalization (GHE) and CLAHE were applied on the AWiFS imagery are given in Table 5.5.

**Table 5.5 Comparison of the quality parameters obtained by employing GHE
and CLAHE techniques on AWiFS image**

Band Description	Technique Used	Mean Square Error	PSNR (dB)
Green	GHE	1.1575×10^9	5.6942
	CLAHE	4.2124×10^7	20.0842
Red	GHE	1.3078×10^9	5.1640
	CLAHE	4.964×10^7	19.3712
NIR	GHE	1.3221×10^9	5.1168
	CLAHE	5.8267×10^7	18.6753
SWIR	GHE	1.3631×10^9	4.9841
	CLAHE	5.6951×10^7	18.7744

It can be easily seen that the images reconstructed using CLAHE technique have a lesser mean square error and a higher PSNR value, thereby resulting in better image quality.

CHAPTER 6 - CONCLUSIONS

CLAHE technique has not been commonly used as far as satellite imagery is concerned. However, it has the capability of efficient feature extraction like shorelines and clouds at lower wavelengths. The CLAHE technique can also be employed in the NIR and SWIR bands of medium and high resolution multispectral datasets to prominently identify wetlands from a given area without calculating the NDWI (Normalized Difference Water Index).

This technique is also efficient in delineating water boundaries in the NIR region of the multispectral image as observed in the results. The use of this technique in the NIR region of high resolution sensors like Sentinel 2A MSI and Landsat 8 OLI can also enhance inland water bodies and can provide the interpreter an idea of sedimented/degraded waters owing to changes in the tone after image enhancement.

It is able to produce images where noise content of an image is not excessively enhanced but the image is provided with sufficient enhancement so as to visualize various features within the image using the CLAHE technique. The CLAHE technique reconstructs images better at shorter wavelengths, but feature extraction is better in the NIR region (especially water bodies and wetlands). The CLAHE technique provides better contrast at longer wavelengths than shorter wavelengths.

The CLAHE technique can thereby be used as a substitute for ordinary histogram equalization for objective enhancement of multispectral satellite imagery.

REFERENCES

[1] Khairunnisa Hasikin, Nor Ashidi Mat Isa, "Enhancement of the low contrast image using fuzzy set theory", 14th International Conference on Modelling and Simulation, 2012

[2] P. Kannan, S. Deepa, R. Ramakrishnan, "Contrast Enhancement of Sports Images Using Two Comparative Approaches", American Journal of Intelligent Systems 2012, 2(6): 141-147 DOI: 10.5923/j.ajis.20120206.01

[3] Shweta Narnaware, Roshni Khedgaonkar, "Image Enhancement using Artificial Neural Network and Fuzzy Logic", IEEE Sponsored 2nd International Conference on Innovations in Information Embedded and Communication Systems ICIIECS'15

[4] M.Selvi, Aloysius George, "FBFET: Fuzzy Based Fingerprint Enhancement Technique based on Adaptive Thresholding", IEEE – 31661, 4th ICCCNT - 2013 July 4 - 6, 2013, Tiruchengode, India

[5] Tarun Mahashwari, Amit Asthana, "Image Enhancement Using Fuzzy Technique", IJRREST, International Journal of Research Review in Engineering Science & Technology (Issn 2278–6643), Volume-2, Issue-2, June-2013

[6] Sargun and Shashi B. Rana, "A Review of Medical Image Enhancement Techniques for Image Processing", International Journal of Current Engineering and Technology, E-ISSN 2277 – 4106, P-ISSN 2347 – 5161, 2015

[7] Rajesh Garg, Bhawna Mittal, Sheetal Garg, "Histogram Equalization Techniques for Image Enhancement", IJECT Vol. 2, Issue 1, March 2011ISSN, ISSN: 2230-7109(Online) | ISSN: 2230-9543(Print)

[8] Md. Foisal Hossain, Mohammad Reza Alsharif, "Image Enhancement Based on Logarithmic Transform Coefficient and Adaptive Histogram Equalization", 2007 International Conference on Convergence Information Technology, IEEE 2007.

[9] J. Alex Stark "Adaptive Image Contrast Enhancement Using Generalizations of Histogram Equalization", IEEE Transactions on Image Processing, Vol. 9, No. 5, May 2000.

[10] Rafael C. Gonzalez, Richard E. Woods, "Digital Image Processing", 2nd edition, Prentice Hall, 2002.

[11] A. K. Jain, "Fundamentals of Digital Image Processing". Englewood Cliffs, NJ: Prentice-Hall, 1991.

[12] M. Abdullah-Al-Wadud, Md. Hasanul Kabir, M. Ali Akber Dewan, Oksam Chae, "A dynamic histogram equalization for image contrast enhancement", IEEE Transactions. Consumer Electron., vol. 53, no. 2, pp. 593- 600, May 2007.

[13] K. R. Castleman, "Digital Image Processing". Englewood Cliffs, NJ: Prentice-Hall, 1979.

[14] D. J. Ketcham, R. Lowe, and W. Weber, "Real-time enhancement techniques," Seminar on Image Processing, Hughes Aircraft, 1976,

[15] R. Hummel, "Image enhancement by histogram transformation," Comput. Graph. Image Processing, vol. 6, pp. 184-195, 1977.

[16] S. M. Pizer, "An automatic intensity mapping for the display of CT scans and other images," in Proc. VIIth Int. Meet. Inform. Process- ing in Med. Imaging, 1983, pp. 276-309.

[17] D. J. Ketcham, R. W. Lowe & J. W. Weber: "Image enhancement techniques for cockpit displays" Tech. rep., Hughes Aircraft. 1974

[18] John B. Zimmerman, Stephen M. Pizer, Edward V. Staab, J. Randolph Perry, William Mccartney, Bradley C. Brenton, "An Evaluation of the Effectiveness of Adaptive Histogram Equalization for Contrast Enhancement", IEEE Transactions On Medical Imaging, Vol. 7. No. 4, pp. 304-312, December 1988.

[19] S. M. Pizer, J. D. Austin, J. R. Perry, and J. B. Zimmerman, "Adaptive histogram equalization for automatic contrast enhancement of medical images," in Proc. 4th Int. Conf Picture Archiving and Commun. Syst. (PACS IV) for Med. Appl., SPIE, 1986, vol. 626.

[20] S. M. Pizer, E. P. Amburn, J. D. Austin, R. Cromartie, A. Geselowitz, T. Greer, B. ter Haar Romeny, 1. B. Zimmerman, and K. Zuiderveld, "Adaptive

histogram equalization and its variations," Compur. Vision, Graph. Image Processing, vol. 39, pp. 355-368, 1987

[21] Zuiderveld, Karel. "Contrast Limited Adaptive Histograph Equalization." Graphic Gems IV. San Diego: Academic Press Professional, 1994. 474–485.

[22] Jacob Lucas, Brandoch Calef, Keith Knox, "Image Enhancement for Astronomical Scenes", 2013

YOUR KNOWLEDGE HAS VALUE

- We will publish your bachelor's and
  master's thesis, essays and papers

- Your own eBook and book -
  sold worldwide in all relevant shops

- Earn money with each sale

Upload your text at www.GRIN.com
and publish for free